Alfred Wingate Craven

Answer of A. W. Craven, Chief Engineer Croton Aqueduct,

to charges made by Fernando Wood, Mayor. New York, July 31, 1860

Alfred Wingate Craven

Answer of A. W. Craven, Chief Engineer Croton Aqueduct,
to charges made by Fernando Wood, Mayor. New York, July 31, 1860

ISBN/EAN: 9783337422943

Printed in Europe, USA, Canada, Australia, Japan

Cover: Foto ©berggeist007 / pixelio.de

More available books at **www.hansebooks.com**

ANSWER

OF

A. W. CRAVEN,

CHIEF ENGINEER CROTON AQUEDUCT,

TO

CHARGES

MADE BY

FERNANDO WOOD,

MAYOR.

NEW YORK, July 31, 1860.

NEW YORK:

BAKER & GODWIN, PRINTERS,

PRINTING-HOUSE SQUARE, OPPOSITE CITY HALL,

1860.

To the

SPECIAL COMMITTEE OF THE

BOARD OF ALDERMEN:

GENTLEMEN,—

A message was transmitted to the Board of Aldermen, by the Mayor, on the 11th instant, in which he says :—

"In consequence of serious disagreements and insubordination, I hereby remove Alfred W. Craven from the office of Engineer of the Croton Aqueduct Board, and Thomas B. Tappen from the office of Assistant Commissioner of the same Board."

In the use of this language, the Mayor appeared to forget that, by the Charter, he had no power to remove an officer without the express consent of the Board of Aldermen, nor without cause. The message proposed to remove me without the consent of your Board, and without any statement of *facts*, enabling your Board to determine whether the "serious disagreements and insubordination" existed in fact, or whether, if existing, they constituted *causes* for removal.

This communication having failed to receive the approval of your Board, the Mayor undertook to correct his error, and, on the 18th inst., transmitted another message, in which he states that he has "the honor of submitting some additional reasons for these removals," and that he is " confident they will meet with the prompt endorsement of the Board."

These two messages have been referred by the Board to you as a special committee, and I have to thank you for your courtesy in granting my urgent request for a full investigation.

Although the assurances of my fellow citizens, constantly volunteered in numerous gratifying expressions, might seem to warrant my treating the accusations against me with disdain, I deem that this very confidence, so unreserved and so kindly exhibited, should be justified and rewarded by a clear and perfect statement of my case. I cannot but feel that such entire trust in my integrity when it is thus assailed, demands from me, in simple gratitude, that I should show it is not misplaced.

In doing so, I must tax the patience of all who for the sake of truth will read or listen to this paper. It is well known that a falsehood can be told in three words, which it will take three pages to disprove, and that this is eminently so where professional and technical matters are concerned.

I will examine each of the Mayor's messages fully and in detail. In order to do so fairly and frankly, I will state the charges against me with a directness which he has failed to attain, and will unfold the covert insinuations with which they are accompanied. The issue is not simply the retention of my place—that is but a trifle in the balance—the issue is one of character—of private and official integrity, and therefore, to me, of life. If I had been assailed by a private citizen, I might have given the assault such a degree of consequence as his personal character would demand. But when my assailant happens to be the Mayor of the city, and, from his public position, embodies his attack in official documents, to remain on the records of the city forever, I cannot let them pass in silence or contempt, but must expose and confute these calumnies, not only for my own sake, but for the sake of my friends, and of all men who appreciate private or professional character.

CHARGES OF THE MAYOR.

The charges in the second message of the Mayor, are four in number.

First. That the work done, and the work remaining to be done upon the new Reservoir, now in course of construction, largely exceeds the original estimate of such work made by me as Chief Engineer; that the quantities of such work in various items differ from the estimate—some being largely increased above, and some very largely decreased below the original estimate; that wherever these changes have occurred, they have proved very beneficial to the contractors, and largely increased the cost of the work to the city, with the covert insinuation that the changes were made by me, because they would be beneficial to the contractors, and because I was acting in their interests, and not in the interest of the city; that I have made a mistake in awarding the contract to the present contractors, whereby the city will suffer loss; that I had made expensive and peculiar changes in the plans of the work, whereby its cost is increased; that I have been wanting in ability, industry, and integrity, whereby the interests of the city have been sacrificed; and that I am disqualified for my position, and some other person more competent should be appointed in my place.

Second. That, in the work of paving Forty-ninth Street from Third to Lexington Avenue, with trap block, the Engineer did not exercise proper vigilance; that the materials employed were not in accordance with the contract, but that an inferior quality of stone was used; and also that the work was performed in so negligent a manner as to cause it to settle in many cases, immediately after its completion; that the Croton Board had notice of the defective character of the materials employed in the work, but, notwithstanding, gave a certificate of its completion and acceptance.

Third. That an error of eight inches in the maximum was made in the grade of a sewer now in progress in Fortieth Street, Ninth Avenue, Thirty-ninth Street, and Eighth Avenue, which error may involve the city in litigation and loss, and which throws a serious responsibility on the Engineer of the Board, showing the want of the constant care and entire accuracy demanded of one to whose professional skill is entrusted the construction of important public works.

Fourth. That John Cornwell and William Conboy, employes of the Department, who had a private job for the excavation of a cellar at Yorkville, were accused before the Croton Aqueduct Board, in December, 1859, by John Chappel and three others, to the effect that Cornwell had furnished powder, fuse, and tools belonging to the Department, to aid in the performance of this private contract, and had allowed men in the service of the Department, and under his supervision, to work upon this cellar, and had fraudulently returned to the Department the time the men were thus engaged, thus causing them to be paid by the Department for labor performed on private work; that notwithstanding the admission of Conboy himself that he borrowed powder, fuse, and tools, from Cornwell, both of these persons are yet retained in the service of the Department; that Mr. Tappen, the then acting President, and myself, the Chief Engineer, ought to have had them indicted for converting the public property to their own use, and had them excluded, by their conviction, from ever receiving or holding any office under the city charter; that we have upheld them in despoiling the city—refused to invoke the most decided vindication of the law against them, and, by keeping them in service, have subverted " the good faith and honesty" of the Department.

Such is the substance of the reasons for which you are called to assist in removing me from my position as Chief Engineer of the Croton Aqueduct Board—a position which I have held for the last eleven years without having, so far as I am aware, ever incurred the slightest suspicion of dereliction in duty, of inability to discharge the responsible labors imposed upon me, or of want of fidelity to the public interests.

I shall answer them in the order in which they are above referred to, and shall in doing so examine somewhat minutely the various items of each charge.

RESERVOIR.

First,—As to the Reservoir.

Under this head, in order to have the whole subject thoroughly understood both by you and the public, I am obliged to go into details which would be unnecessary if I were addressing a body of engineers.

It is necessary then to understand what this Reservoir is to be—what the land upon which it is constructed originally was —for what purpose the estimates of the various kinds of work upon it were made—and in what manner we arrived at the supposed quantities of work to be done.

The object of this construction is to obtain space for impounding or storing a sufficient quantity of water within the city limits, and in immediate connection with the street mains, to supply the inhabitants with water during such time as any repairs to the main aqueduct or conduit of masonry make it necessary to shut off the usual daily flow into the city. The present Reservoirs are insufficient for this purpose. The amount stored in them is so small compared with the daily consumption, that we are unable to keep the water shut off during a greater period than seventy hours at a time, whatever repairs may be necessary—while the condition of some parts of the aqueduct is such, that the necessity for more thorough examination and repair than can be done in that period, is steadily becoming more pressing. This subject engaged my attention almost immediately after my original appointment as Engineer, and foreseeing the importance of the early completion of this work, I urged forward, as far as lay in my power, the legislative steps necessary for its commencement. The property selected for its site was not, however, obtained until 1856, and the appropriation for its construction was not made by the legislature until 1857. The site lies between 85th and 97th streets, and between 5th and 8th avenues, embracing an area of 106 acres, and although situated at considerable height above the level of tide water, the centre of it was a natural basin surrounded by land and rocks of various degrees of elevation. This basin was a swamp or morass, covered more or less with water, and rendered difficult of accurate examination by the thicket which covered the surface and filled the soil with its matted roots. The charter of the city required that this, like all other similar work, should be done by contract, and be awarded to the lowest bidder without any distinction—and the established practice of the Federal, State, and City Governments does not permit that work of this kind should be let or taken *in solido*, or by the lump. The practice requires that an *estimate* should be formed of the relative quantity of *each item* or *kind of work and material* to be employed in the entire construction, and that contractors should bid a certain *rate* or *price for each item* or *kind*. Such rates or prices, multiplied by the estimated quantities, would give the total result of each bid—and a comparison of such totals would determine who was by the law, the lowest bidder. The undertaking of the work was not dependent on its cost. The Reservoir was *a necessity*, and would have been ordered by the city even had the original estimate of its cost been double what it was. But even had the undertaking been dependent on its cost, so long as the government was not deceived by my estimate of the total cost (*which I shall clearly show hereafter*), minute exactitude in the relative proportions of the different quantities making this total cost, was not indispensably necessary for the purpose of letting the work. It was desirable that they should be as exact as it was possible to determine; but as each contractor would bid upon the same estimate of quantities, and the real object was simply to ascertain, in obedience to the charter, who was the lowest bidder upon a supposed basis, it is clear that when the estimate was made, the bidders stood equal and upon the same footing with each other.

But how was even a probable estimate of each kind or item of work and materials required for the total construction, to be ascertained? The *surface* of the entire area of 106 acres could be seen with the naked eye. In many cases the surface gives indications of what will be found underneath. But this area of land was peculiar, as I have before described it—and who could tell *what would be found beneath its surface?* how much solid rock? what kinds of rock? how much earth? what kinds of earth? how much of the materials, either of earth or rock, would be fit to use in the construction? Engineers have no more power of subterranean insight than ordinary men. We did the only thing practicable in such cases. We examined with *unusual care* the entire surface, and by excavations in some places, and the use of iron "*sounding rods*" at very short distances apart over the entire area, we arrived at an approximate result. Our borings with the rod were carried down as far as possible, sometimes reaching to the depth of thirty feet; but these soundings could not always be relied upon with entire certainty. In some cases the rod might strike a rock which would indicate the solid ledge, but which, upon the development of the work, would prove to be simply a large boulder, or else an upward projecting point far above the general level of the ledge. So also with the kinds of rock and the kinds of earth, no data could be given which would be entirely reliable. The officer in immediate charge of these examinations was the present efficient engineer of the work, Captain George S. Greene, late of the U. S. Army, and for some years Professor of Mathematics and Engineering at West Point Academy. A more thorough investigator or more upright man can nowhere be found. And I assert confidently that a closer approximation to the real quantities, as developed in the construction of this work, could not have been made on this piece of ground, unless at an expense which would be entirely unwarrantable. We did not therefore expect to obtain a result showing the *actual* number of yards of excavation, whether of earth or of rock, or the *precise character* and proportions of the materials which the site of the proposed reservoir contained. We simply expected to obtain data which would enable the contractors competing for the work to form some judgment of its requirements and its cost, and thus put them on an equal footing, and further to enable me to give to our Board my own views of its probable total cost. The examinations resulted in the following estimate of quantities :—I quote from the specifications.

" (31.) Quantities estimated for the construction of the work, on which the comparisons of bids will be made.

* E. 792,519 cubic yards of excavation, composed as follows, viz: 63,304 cubic yards of soil and other materials not suitable for the construction of embankment, to be placed in spoil banks, within the space mentioned in specification (1), (of which 17,362 cubic yards will be placed on the outer slopes of the banks); and 25,000 cubic yards of excavation from the puddle ditches, to be placed in embankment or spoil bank; 704,215 cubic yards of excavation for puddle and embankments.

F. 284,894 cubic yards of rock excavation.

G. 78,988 cubic yards of puddle in the ditches and embankment.

H. 642,589 cubic yards of embankment, exclusive of the puddle bank.

I. 20,938 cubic yards of broken stone lining of inner slopes.

K. 47,111 cubic yards of stone paving or slope wall.

K². 2,009 cubic yards of paving laid in cement.

O. 555 cubic yards of brick wall, laid in cement.

P. 833 cubic yards of concrete on the middle bank.

Q. 54,000 square feet (superficial measure) of sheet piling."

At the same time, being aware of the uncertainty of any quantitative estimate in a case of this description, and being desirous on the one hand that the city should be guarded against extra claims, and that on the other hand, no competitor for the work should be misled, the following clause was inserted in the proposals, and immediately following the estimate of quantities :—

"Other quantities, dependent on developments in the progress of the work, are not estimated. The foregoing are the quantities which have been estimated, approximately, for the construction of the work; they form, however, no part of the contract, and persons bidding are cautioned, that the Croton Aqueduct Board do not hold themselves responsible that any of them shall strictly obtain in the construction of the work, and the contractors are required to examine the plans and the ground, and to judge for themselves, as well of the quantity, as the distance of haul, and other circumstances affecting the cost of the work, and to make their bids for each item independent.of others, so far as relative quantities are concerned."

The above clause was also included in the contract and specifications.

The proposals having been advertised, bids were received on all the items of work and materials, from seventeen different individuals or firms. The bids were so much per foot or per yard for each item, and multiplied by the estimated quantities resulted in totals varying from $614,001 97 to $1,914,730 46. And this extraordinary difference in the *totals* was not greater than the rates or prices for which the different competitors proposed to perform the *various items* of the work—showing conclusively, what in practice is well known, that the price or rate at which a contractor is willing to do one portion of a large work, is no evidence that such rate or price is considered by him to be remunerative. They frequently propose to lose by some portions of the work, and to make up the loss by the price or rate which they will obtain for other portions.

The contract was awarded in the summer of 1857, by our full Board and the Comptroller of the city, to Fairchild & Co.; the law requiring such award to them as the lowest bidders complying with the ordinances prescribing the form of making bids. The total of their bid was $632,473 33 ; that of the next highest bidders (Cummings & Co.) was $729,807 25. The contract was not confirmed by the Common Council until the spring of 1858.

With these general explanations, I now proceed to the first subdivision of the Mayor's general charge in regard to the Reservoir. On the 2d of May, we furnished him at his request, a table showing the quantities of work already done, and comparing them with the estimate; but without asking from us one word of explanation, he sends in his message of July 18th,

in which he states that the following remarkable discrepancies
appear between the work as done and as estimated.

INCREASE OF EXCAVATION.

First. "That the excavation has been increased from thirty
to sixty thousand cubic yards, or 7 per cent."

I have shown you the inherent difficulty in forming a close
quantitative estimate for such a work on such ground, and the
strong probability that the quantities would not be verified by
the actual progress of the work. I have shown you that my
experience enabled me to foresee these contingencies, and that
I did foresee and did provide against extra claims against the
city, by the warning to the contractors and the express stipula-
tions of the contract itself. I will now show you, under each
head, the special cause for such discrepancies, and I shall then
leave you to determine whether it is fair or manly to say of
them, with obvious obliquity, *and without asking a word of
explanation*, that they are "remarkable."

The exact quantity of excavation is even now a matter of
conjecture; but whether it will involve an increase of 7 per
cent. or not, it is nevertheless true, that it will be considerably
in excess of the amount as originally estimated. The reasons for such increase are these :

REASONS FOR INCREASE OF EXCAVATION.

First. When the work was commenced, the more depressed
portions of the site of the Reservoir were very wet. To render it
fit to receive the embankments, and also to enable us to pro-
cure by excavation, proper material for those embankments, it
was necessary to make large ditches through the site of the
Reservoir to draw off the water.

This, of course, increased the excavation ; that is to say, increased it to the extent of the number of cubic yards of excavation which each ditch contained.

Second. The material thus excavated, instead of being carried directly into embankments, was necessarily collected in
temporary places of deposit. When the ground was sufficiently drained to permit the construction of embankment, it

was necessary to remove the material from these temporary banks to permanent embankments, thus involving a double excavation. The work upon the Reservoir was commenced in the spring of the year 1858, when the ground was wet from the melting of the snow and from the spring rains; and far greater difficulty was experienced on this account than would have been the case, if the contractors could have entered upon it early in September, 1857, as was anticipated at the time the estimate was made and the contract awarded.

The Common Council, however, did not confirm the award of the contract soon enough to enable the contractors to commence the work at that time.

Third. Upon excavating, it was found that in many places the earth was not of proper quality for embankment, and the contractors were obliged to dig much lower in the site of the bottom of the Reservoir than was anticipated, to obtain earth which was suitable.

The contractors, it will be seen by a reference to the first section of the contract, were bound to take their materials for the construction of the Reservoir from the ground, if they could be obtained; and if they could not be obtained upon the ground, they were entitled to charge, for such of the materials as they procured *from other sources*, a price *additional* to that agreed upon in the contract for similar kinds of materials found *within* the lines of the Reservoir.

It was, therefore, to the interest of the city, that materials should, if possible, be obtained from the Reservoir grounds, and it was *to prevent an increase* in the price of the materials necessary for embankment, that I directed the increased excavation to be made *within* our grounds.

It was foreseen that this difficulty might arise, and a provision was inserted in the 10th section of the specifications to this effect:

"(10.) The earth, within the working lines of the interior slopes of the basins, will be excavated to the depth of forty feet below the top of the exterior Reservoir banks, *and as much lower as may be required by* the engineer, *to obtain materials for the embankments or for the puddling.* The surface of the bottom of both basins to be worked to the same depth, and to such slopes as the engineer may direct. The exact depth cannot be determined until the surface of the rock is uncovered. Whenever it may be necessary to go below the bottom of the Reservoir, as it shall be established by the engineer, to obtain materials for embankment or for puddle, the space so excavated shall be filled up with such waste materials as the engineer may direct."

Here, it is shown that we were fully authorized by the contract to order an increase in the excavation, should it become necessary for the purpose of obtaining materials for embankment. The necessity did arise, and the excavation was materially increased thereby. The pecuniary advantage accruing to the city (*not the contractors*) by the order to obtain the required quantity of material from *within* the limits of the Reservoir will at once be perceived by referring to the Contract. Section

"(E.) For all excavation, whether hardpan, quicksand, stones, boulders, or otherwise, excepting solid rock which requires blasting, and excepting boulders more than half a cubic yard in capacity, and for disposing the materials according to the specifications in spoil bank, or in refilling excavations made to obtain materials, or embankments, or puddle bank, or puddle, to be measured in excavation, twenty-one (21) cents per cubic yard.

"(H.) For embankments, including the soil one foot in depth on the outer slopes and exclusive of all puddle and all masonry and broken stone on the slopes, in addition to the price paid for excavation in item E, or for furnishing materials in item N, when the engineer may direct, materials to be furnished from outside the Reservoir ground, to be measured in embankment, one (1) cent per cubic yard.

"(N.) For earth to be furnished for embankment, should the engineer direct any to be obtained *beyond* or *outside of the Reservoir* ground, in addition to the price paid for embankment in item H, to be measured in embankment, fifty (50) cents per cubic yard."

That is to say : If the earth for the embankment were furnished from within the Reservoir grounds, the contractors, for excavating and putting it in bank, were to receive 22 cents per cubic yard ; if obtained from other sources, they were to receive 51 cents per cubic yard for it.

4th. The nature of the Rock encountered in the actual working, materially added to the amount of rock excavation as originally estimated. In many places the rock was found to be of a porous, friable character, full of cracks and fissures, and utterly unfit for the base of the puddle trench. This contingency was also provided for.

The 12th section of the specifications in the contract contained this provision :—

"When any fissures, seams, or soft rock, are found within the basin of the Reservoir, or under the banks thereof, the rock is to be excavated to such extent as the engineer may direct, and the space filled with concrete, puddle, earth, or sand, as may be required by the engineer."

It will be apparent from the above, that not only was our Board authorized to carry out these portions of the contract providing for this *method* of prosecuting the excavation, but that my duty, as an engineer, left me no alternative but to avail myself of the *provisions made in the contract for this express purpose.*

Let us now proceed to the consideration of the second specification of the Mayor's charge, under the head of the Reservoir.

INCREASE OF PUDDLE.

2d. "That the puddle has been *increased* by an amonnt of twenty-five or thirty thousand cubic yards, or thirty-seven per cent."

Before entering upon this subject, I will briefly describe what "puddle" is :

A "*puddle trench*" or "*wall*" is a space in the centre of the embankment which surrounds the Reservoir. It is made of mixed materials (earth, clay, and sand) worked together carefully with spades while wet. Properly made, it is impervious to water, and is the barrier on which we rely to prevent the banks from leaking. The embankment alone cannot be depended on for this purpose. Although in some degree watertight, the main function of the embankment, exclusive of the "puddle," is to resist the lateral thrust of the impounded water. Below the base of the embankment, the space for this puddle is carried, in the form of a trench, to the solid rock. Above the base of the embankment it is carried up, or constructed, at the same time with the embankment, and it is then sometimes called a "puddle *wall*." If the embankment rests on, or starts from ordinary earth, and the puddle trench is not carried down to the rock, the water will pass through the earth below the puddle wall, and thus escape. To ensure perfect work, therefore, it is better to carry the puddle trench down to the rock in all cases where it can be reached, within available distance, unless, indeed, the earth on which the embankment rests is in itself impervious to water. In the new Reservoir the water will be enclosed by a puddle wall, resting upon the solid rock throughout its entire length. At no one point, therefore, can the embankment leak.

With these explanations and remarks, I will proceed to the

CAUSES OF INCREASE IN PUDDLE.

1st. The porousness and friability of the subterranean rock encountered in the puddle trenches, which, as I have above shown, caused an increase in the rock excavation, produced necessarily a corresponding increase in the puddle. In many cases the fissures and holes in the rock rendered it utterly unfit for the bed or base of a puddle trench, and the contractors were required, in such cases, to continue their excavation until they reached a bed of solid rock which *was* suitable for this purpose. This, of course, produced both an increase of excavation and of puddle. The deeper the excavation for the puddle trench, the greater the amount of puddle required to fill it.

2d. In sounding the ground for estimates of quantities with the iron rods used for such purposes, an engineer is frequently deceived as to the amount of rock contained in the ground, from the fact of the rod happening to strike upon a boulder or a point of rock which appears to be the main ledge. This occurred in our examinations. Upon excavating, it was found, in many instances, that where solid rock was presumed to exist, the engineer's sounding rod had been stopped by a boulder or by an upward projecting point of rock, and that the main ledge was far below the level indicated by the rod.

In some such instances we were obliged to go ten feet deeper to reach the main ledge of rock. Under the southern bank, we were compelled, by the character of the rock, after reaching the ledge, to excavate through it twenty feet further to get down to a bed which was proper for the puddle The additional excavation, both of earth and rock (and the consequent increase of puddle), owing to these causes, under the western bank through the swamp, was exceedingly great, the character of the natural earth requiring its entire removal for a width equal to half the width of the base of the embankment. This excavation extended down to the rock, and the space was filled up with puddle—a proper and justifiable arrangement of materials.

By section (7) of the specifications and contract, you will see that the contingency of increase in this particular also, was foreseen and provided for.

"(7.) Puddle ditches are to be excavated to the rock under the centre of all the embankments, where the rock is not lower than forty-six feet below the top of the exterior Reservoir bank: and *in any loose, broken, soft, or porous rock, the ditches will be excavated to such depth as the engineer may direct.*"

So much for increase of puddle.

DECREASE IN EMBANKMENT.

The 3d subdivision of the Mayor's 1st charge is:—

" That the embankment has been decreased by an amount of 215,565 cubic yards, or one-third less."

The reasons for this decrease are these:—

1st. From the apparent nature of the ground it was in many places uncertain how far we should have to go, below the natural surface, to find good material for the bed of an embankment. A large allowance was made for this, and provision made in the specifications for decrease, should it be found advisable. As our work was developed, it was found that at certain parts of the work we should not be obliged to go to such depth for a proper bed as we anticipated, and for which depth we very properly had made an estimate. Less embankment was therefore necessary—but for the same reason less excavation was required; first, the excavation required to go down to the proper bed for the embankment; and, second, the excavation necessary to obtain materials to make the embankment, had it (the embankment) not been thus decreased. As a yard of excavated earth will not make a yard of compacted embankment, the saving or decrease of excavation by this development was more than twofold the amount of decrease in the embankment.

Thus, a large amount was in reality saved to the city; and if the price for excavation in the contract was a remunerative one (which, notwithstanding the assumption of the*Mayor, does not appear), this development or so-called "*change*" was greatly against the interests of the contractor.

2d. Another reason for decrease is in the embankment yet to be done, and it is this : A road now contemplated by the Central Park Commissioners, around the Reservoir, enables us to fill against the outer slopes of our banks with waste materials out to a level with their road, instead of covering these slopes from top to base with a foot of selected soil for a turf protection to said slopes. The amount of this soil was originally included in the comparative estimate, as embankment.

STONE PAVING AND PAVING IN CEMENT.

The 4th and 5th subdivisions of the Mayor's first charge are as follows :

"*Fourth.*—That the stone paving has been decreased by an amount of 46,941 cubic yards, or $\frac{1}{210}$ only of the original amount.

"*Fifth.*—That the paving in cement has been increased by an amount of 18,867 cubic yards, or more than nine times greater."

REASONS.

Inasmuch as the decrease in the dry slope wall, or what is designated in the contract as the "*stone paving,*" arose from the same cause which increased the amount of the paving in cement, I shall answer these two subdivisions together.

The increase in the paving in cement will be to a still greater amount than mentioned by the Mayor ; to wit, it will be increased to the same extent as the dry or stone paving is decreased. To explain :

It will be seen by reference to section 19 of the specifications of the contract, that it is provided that,

" The interior slopes of all the banks will be covered with 8 inches in thickness of stone broken to pass through a 2-inch ring ; on this will be laid the paving, 18 inches in thickness, of a single course of stones set on the edge at right angles with the slope, *laid dry,* and well wedged with pinners. The stones to be laid in such manner as may be *directed by the Engineer,* the base of the paving to extend *to the rock, or to such depth as the Engineer may direct ;* the stones to be sound, and *of proper shape to make neat and compact work.*"

2

The original intention was, therefore, to build a dry or rip-rap wall on the interior slopes of the embankment. The estimate of the quantity of paving required for these slopes was 47,111 cubic yards (see Estimate). This estimate will correspond almost exactly with the quantity which will be actually put up when the contract is finished. *The quantity of stone paving has not been increased or diminished.* It is simply laid up in cement instead of being laid dry. The reasons and authority therefor I will now show.

It was found, as the excavation progressed and the character of the rock within the lines of the Reservoir was developed, that the stone was *unsuitable for a dry wall.* About 175 cubic yards of this wall *was* laid dry, and the stone was found to be so unfit for a serviceable wall, that the Board directed the contractor to take it down.

The question therefore arose whether the contractor should be allowed to furnish suitable material from *other* places *outside* of our lines, or whether the stone should be laid in *cement.* The contract contained full provisions to meet this contingency also. See contract (section 17).

"Should the excavation *within* the outer slopes of the exterior Reservoir banks not furnish sufficient material of a *suitable* quality for puddling and embankments, in the opinion of the Engineer, then the contractor will be required to furnish enough material to supply the deficiency, and of such quality as may be approved by the Engineer."

By section K of the Specifications it is provided, that the contractor shall receive, "for paving or constructing slope wall, *the stone to be taken from the rock excavated,* and for labor in constructing the same, $1 25 per cubic yard."

By section K² it is provided, that for the same stone, "laid in cement mortar, including all labor and material to be measured in the work, the contractor shall receive $4 00 per cubic yard."

By section L, it is provided that the contractors shall receive, " for stone to be furnished for the paving, in case the *stone should not be deemed suitable for paving by the Engineer, in addition* to the price above stated in item K, $3 per cubic yard."

The price, therefore, for a *dry* wall to be paid to the contractors, in case they furnished the material from *outside* of the Reservoir lines was by the contract $4 25 *per cubic yard ;* while the price to be paid to them, in case they used, for the wall, the stones found *within* the lines of the reservoir, *laid in hydraulic mortar*, was as above shown, *only $4 per cubic yard;* and assuming that the total quantity of slope wall when finished will correspond with the quantity estimated, to wit, 47,111 cubic yards, we shall have the whole laid up in *hydraulic mortar*, for $11,444 42 *less* than it would cost if laid *dry* with any stone available outside the reservoir ground.

The whole of this subject was most carefully and thoroughly examined by myself personally, and at different times by myself in conjunction with my colleagues in the Croton Board. The ground was visited, the stone examined, and the question, engaging, as it did, our earnest attention, was fully discussed in all its bearings.

Having become satisfied that it would be utterly impossible, within any reasonable expenditure, to construct a suitable dry wall out of the rock found within the lines of the reservoir, and the contract providing that the contractors should receive less for paving in cement-mortar with the stone found within the lines, than for building a dry wall out of stone procured by them elsewhere, at a meeting of the Croton Aqueduct Board, held on the 20th day of May, 1859, at which my then associates, Messrs. Myndert Van Schaick and Theodore R. De Forest, were present, I introduced the following preamble and resolution, which were then unanimously adopted :—

" *Whereas*, It has been found (by the Chief Engineer) that the excavation within the lines of the new Reservoir does not furnish stone of sufficiently good quality to make a suitable facing for the interior slopes, if laid dry, but that the same stone will make a good slope wall if laid in hydraulic cement ; therefore,

" *Resolved*, That the said slope-wall be laid in cement to such extent as the nature of the stone may make it advisable ; such change, and *such character of the work being provided for in the contract.*"

It is proper to state in this connection, that a wall of stone laid in hydraulic mortar, in the manner in which this work is done, is better for the purposes for which it was designed, than any dry wall which could be constructed of any stone whatever, of equal dimensions.

The dry wall was originally selected as a matter of economy; and if the stone had proved good, a dry wall constructed of it would have been sufficiently substantial in the outset, though liable to require earlier repairs. When I found that the stone excavated would not answer the purpose, I abandoned the dry wall, and resorted to the substitute (*already provided for in the contract*) which involved the least expense to the city.

In referring above to my colleagues, I would not be understood as wishing to divide the responsibility of this substitution. As the professional adviser of our Board, I gave my opinion—and gave it decidedly ; and in this, as well as in all points touching the Reservoir, I am not only willing, but anxious, to assume the entire responsibility. Everything that has been done, has been solely for the perfection of the work, and *nothing has been done which was not a clearly-expressed part of the contract.* Even had we not been so entirely fortified by the forethought shown, and provisions made in the contract, I would have recommended work for which there was no provision, rather than that the Reservoir should be imperfect. Nay more—even could I have imagined that any one existed who could impute to me unworthy motives for my action, I would have advised or directed all that has been done under the contract without reference to the question whether or not it would be beneficial to the contractors ; and for this simple reason—that I owed it to the city to do their work honestly and well, without being deterred by the risk of personal calumny or contemptible suspicion.

There has been in fact therefore *no decrease in the stone paving ; the same amount* of stone paving will be laid as estimated—but instead of being laid dry, it is laid in hydraulic mortar for the reasons above stated.

INCREASE OF CONCRETE.

The 6th subdivision of the Mayor's first charge is : "That the concrete masonry has been increased by an amount of 33,568 cubic yards, or more than forty times the quantity originally estimated."

The reason for this increase was the character of the rock hereinbefore described, as developed in the depths of the excavation for the puddle trenches. It was found so full of fissures and seams, even after blasting the porous and disintegrated portion away, that it was deemed essentially necessary to cover it with a layer of concrete, as a bed or base for the puddle.

It was supposed there would be more or less of these fissures and seams encountered, but it was impossible to divine their number or extent, and therefore impossible to make any accurate estimate in respect to the quantity of concrete necessary to fill them.

Anticipating, however, that concrete might probably be required, the following provision was inserted in the contract:

"SECTION 12. Where any fissures, seams, or soft rock are found within the basin of the Reservoir, *or under the banks thereof*, the rock is to be excavated to such extent as the Engineer may direct, and the space filled with *concrete*, puddle, earth, or sand, as may be required by the Engineer."

As it could not be known whether it would be necessary to lay a bed of concrete for the puddle, and if so, to what extent, an estimate was made simply of the concrete which would be required on the surface of the middle or dividing embankment,—833 cubic yards, which is just the amount which will be required there.

It will be seen from the above, that the variations from the estimated relative quantities by way of increase or diminution in every item of work alluded to in the Mayor's message, have resulted either from the nature of the ground as developed, or the character of the materials as excavated; that the estimate itself was not, and was not intended to be, an accurate statement of quantities, but the best approximation to accuracy which, under the circumstances, could be obtained. As an Engineer, I do not hesitate to declare that the estimate was thus accurate in all its parts. The only *apparent* change of plan from the requirements of the contract, was the substitution of a wall laid in hydraulic cement, for a dry wall on the interior slope of the Reservoir. I have shown that this "change" was foreseen and provided for, and was induced by the nature of the rock encountered, and by the desire on my part to avoid the necessity of procuring materials for a dry wall from outside the Reservoir lines, at a greater cost. I have also shown that this "*change*" (if that can be called a change which is provided for in the contract) was concurred in by the Croton Board unanimously.

It may, however, be inquired what authority had the Croton Board to authorize this substitution. The answer will be found in the following extract from the contract itself, which not only authorizes all that we have done, but would have authorized an *actual change* of *plan*, had the developments of the work made a change necessary for its perfection.

"And said parties of the second part further agree, that said Croton Aqueduct Board, or the engineer, may make alterations in plan, form, or dimension of said work, either before or after the commencement of the construction. If such alterations diminish the quantity of work to be done, they shall not constitute a claim for damages, or for anticipated profits on the work that may be so dispensed with. If they increase the amount of work, such increase shall be paid for only according to the quantity actually done, and at the price established for similar work under this contract; and the contract will be regarded as completely executed and fulfilled, when the work required by the altered plans is done and paid for, as herein provided.

And it is further agreed, that if the work shall be increased by the enlargement of any part of the same, or by any contingent work which the engineer may deem necessary to facilitate the execution, or *render the work in any particular conformable to local circumstances*, or which may be deemed by the engineer necessary for perfecting the work, beyond what is provided for in this contract and specifications, such increase shall be paid for at the same rate as similar work is herein contracted to be paid for; and if such work is not similar to work herein contracted for, it shall be paid for as an extra item, at a price to be agreed upon previously to the commencement of such extra work."

It was under this authority that the Board acted in making this "change;" an authority which was reserved to them by the contract for the *express purpose of meeting such an exigency.*

CHANGES BENEFICIAL TO CONTRACTORS.

But it is alleged against me, that "whenever these changes have been made, they have proved very beneficial to the contractors, and, in most instances, have largely increased the amount which the city must pay for the work."

What is the proof? The Mayor admits his ignorance of the subject. He says: "It requires either the professional knowledge of an engineer, or the experience of a contractor acquainted with such work, to determine the actual value of the different kinds of work required for the Reservoir." And he adds : "Not having this knowledge, I have availed myself of the opinions of the most skillful contractors." And where are their opinions ? He finds them in the bids of Cumming & Co. *for those particular items of the work which happen to be lower* than the bids on the same items by Fairchild & Co. The *total* bid of Cummings & Co., was the next above to that of Fairchild & Co. What authorizes him to say that the bids of Fairchild & Co., being in some instances higher than those of Cumming & Co., shew them to be remunerative ? I have shewn to you that by universal experience, the bids for special items of a work are no evidence that such bids are remunerative; on the contrary, that bids on certain items are frequently made at a known loss. *Why select the bids of Cumming & Co., and ignore the bids of the sixteen other contractors ?*

Take the items quoted by the Mayor :

Excavation has been *increased* beyond the estimate. Fairchild & Co. receive 21 cents. The bids for this item varied from 12 to 60 cents, one being at 12 cents, two at 20 cents, and Cumming & Co. being 22. Who can tell, testing the matter by such bids, whether 21 cents is remunerative or not ? or, to use the language of the Mayor, "very beneficial to the contractors." By the test which he has adopted, of selecting the *bid* of Cumming & Co. as a criterion, Fairchild & Co. must be making the excavation *at a loss.*

Stone-paving has been *decreased* below the estimate. Fairchild & Co. receive 125 cents. The bids on this item varied from 90 to 750 cents, the bid of Cumming & Co. being 135. Who can say that it was " beneficial to the contractor " to have this item *decreased* when other bidders were ready to do it at 90 cents ?

Paving in cement has been increased above the estimate. Fairchild & Co. receive 400 cents. The bids on this item varied from 250 to 850 cents; that of Cumming & Co. being 350 cents. Who can say that the *increase* of this item was " beneficial " to the contractors at 400 cents, when other bidders were not willing to do it for double the amount of the compensation which they are to receive ?

The *concrete masonry* has also been *increased*. Fairchild & Co. receiving 500 cents, the other bidders varying from 298 to 800 cents. Four only out of the seventeen bidders were willing to do it for less than 500 cents. Eleven bidders asked from 500 to 800 cents. Which was the remunerative price? Who can say that it was "beneficial" to Fairchild & Co to do this item at their price?

But why in this minute catalogue did the Mayor omit another item, to wit, *puddle*, which, he says in another place, has been *increased* beyond the estimate more than 37 per cent? The bids on this item varied from 20 to 150 cents. Fairchild & Co. bidding the *lowest price of all*, viz., 20 cents; Cumming & Co. bidding 30 cents. Was this *increase* beneficial to Fairchild & Co.? By the test adopted by the Mayor it was very prejudicial to them. *Why then did he not mention it?*

The allegation of the Mayor, therefore, that "whenever these *changes* have been made, they have proved very beneficial to the contractors" (admitted by him to be made in total ignorance), is wholly contradicted by the only test by which he seeks to establish the charge.

Independent of this fallacious test, however, I am entirely satisfied from my own observation and experience, that the prices paid Fairchild & Co. for the excavation and the puddle, the quantities of which have from necessity been largely increased, are not remunerative; that the other items of increase, to wit, the paving in cement, and the concrete masonry, considering the strict manner in which the contractors are held to the letter of their contract, are paid for at prices which are only fair and by no means extravagant.

In fact, in the original calculations of the cost of the entire work, made for the use and guidance of the Board before even the proposals were drawn, I set down the probable price for excavation and puddle at a higher figure than those payable under the contract; and the probable prices for slope wall, if laid up in mortar, and for concrete, at about the same figure as is now payable under the contract.

At this point may very properly be introduced a statement of what really was the original estimate of the cost of this

Reservoir, in which the Mayor alleges so gross a mistake was made by the Engineer. The Mayor's charge is so worded as to lead you to the belief that my original estimate of the cost, exclusive of the gate-houses, was some $630,000, and that its actual cost will be $100,000 more than that sum. This, as he well knows, was simply the bid or estimate of Fairchild & Co., made in competing for the work. Now, I am prepared to show you, that my original *estimate* of the probable cost of the whole work, exclusive of the gate houses, made for the guidance of the City Government, was in round numbers $1,025,000, instead of $632,473 24, which was simply the bid of Fairchild & Co. If you will add what the Mayor is pleased to term my "mistakes" to this sum of $632,473 24, you will still find that the total cost of the work will be at least $100,000 less than my estimate, instead of being over $100,000 more.* The reasons that my estimate will now be found to be ample, are, 1st, That my prices on the several items, making up the total, were higher than those *bid* by a majority of the competitors; and, 2d, because to this total, my experience as an Engineer cautioned me to add sufficient to meet all the contingencies which have been already actually developed in the progress of the work, and which I have so minutely described.

* As it may be inferred from the Mayor's statement that the cost of the new Reservoir will be excessive, it is well to make here a comparison which will be understood at a glance:

The present Distributing Reservoir cost $434,551
The present Receiving   "   "   414,500
The new Reservoir now under construction, will cost, including Gate-
   houses, and making full allowance for developments............ 1,000,000
The capacity of the new Reservoir is 5.625 (or more than 5½)
   times greater than that of the present Reservoir, and 40.18
   (or more that 40) times greater than that of the Distribut-
   ing Reservoir. In other words, the cost, per gallon of ca-
   pacity is as follows:
Distributing Reservoir....... .... 2.316 cents.
Present Receiving Reservoir .... .276 of a cent.
New Reservoir .... .119   "

No arguments as to the relative economy of building *large* or *small* Reservoirs, will weaken the fact here shown; that, by comparison, the new Reservoir will be finished at an exceedingly reasonable price, to say the least.

From the facts here stated in regard to the extent of area, and the peculiar character of the ground, no one will have been surprised to learn that the actual quantities have differed from the approximate estimate, while many will be surprised only because they differ so little. The *only point*, then, rests in the allegation of the Mayor, that these differences or variations have, in every case, proved beneficial to the contractors. I have said enough, I trust, to show that this charge, like its predecessors, is wholly unfounded.

MISTAKE IN AWARDING THE CONTRACT.

But, it is said that I made a mistake in awarding the contract to Fairchild & Co., as lower bidders than Cumming & Co.

This is not true. In the first place, the award was not made by me personally, but by the entire Croton Aqueduct Board. In the next place, there can be no "mistake " in an estimate of this kind, where there is not only no assertion or even pretence of perfect exactness, but an actual provision for, and warning against contingent developments. In the third place, the award was *perforce* made in obedience to the charter, which requires work to be let to the lowest bidder.

The Mayor does not pretend to deny that the bid of Fairchild & Co. was lower than that of Cumming & Co., by $97,334 32, nor that the bids were tested by the estimate made for the purpose of ascertaining who was the lowest bidder, while it has been already fully shown no closer approximation to the developed facts than that estimate could have been made, without incurring an expense unwarranted by the circumstances.

RUMORS.

Having fully refuted all that is either openly uttered or covertly insinuated in the *expressed* charges of the Mayor as to the new Reservoir, I beg to call your attention, for a moment, to the extraordinary language with which he concludes this portion of his message.

He says, "In commenting upon this subject, I have abstained from giving currency to any of the *rumors* which have reached me." With what object, let me ask, was this allegation made? Were the rumors which float in the air of the Mayor's office true, or were they false? He does not condescend to inform you. What was their substance? He does not state. From whom did they proceed? and who were engaged in their promulgation? He is silent. Did they come from the vicious and corrupt, whose efforts at extortion I had baffled and defeated? or from some schemer for my office? or from the parasites of power, who hoped for favor, by offering up the incense of my slaughtered reputation? He does not inform you. He merely states that he has "abstained from giving currency to any of the rumors." If he professed to be too honorable to give them currency, why did he not abstain from stating the fact, if *such a fact there be*, of their existence? Was it intended to set the imagination of men at work, and to allow them to give these rumors their form and feature, each according to his own fertility and caprice? I have now been engaged more than twenty-five years in the arduous duties of my profession. For eleven years past I have been known to my fellow-citizens in the responsible position I now hold. Both here and in other places, I have successfully prosecuted important public works, costing millions of dollars, and until I read this message of the Mayor, I never heard of any remarks prejudicial in the slightest degree to my reputation. For my own part, I do not believe that the Mayor has ever heard a whisper to that effect from lips that were honest; and I ask you, and all men, who, having character, know its value, what expression in the English language is applicable to one who, in hunting a man down for his office, attempts in this manner to " *rumor* " *away his reputation?*

PAVING OF 49TH STREET.

I ask your attention now to the second charge of the Mayor, viz. :—

That in the work of paving 49th street from 3d to Lexington Avenue, with trap blocks, the Engineer did not exercise proper vigilance ; that the materials employed were not in accordance with the contract, but that an inferior quality of stone was used, and also that the work was performed in so negligent a manner as to cause it to settle in many cases immediately after its completion ; that the Croton Board had notice of the defective character of the material employed in the work, but notwithstanding, gave a certificate of its completion and acceptance.

What is the proof to sustain this accusation? Nothing more than the unsworn statement of nine persons claiming to be owners of property on the line of the street, and interested in defeating the assessment for the work.

Did the Mayor investigate this charge? Does he believe it to be true? On the contrary, in the same breath in which he makes the charge, he expresses entire ignorance of it; because he says, "*either* the property owners are utterly mistaken in their emphatic condemnation of the work, *or* Mr. Craven has been guilty of gross negligence in its supervision and examination." I will now proceed to show the facts of the case.

I find, upon inquiry of the Water Purveyor, who is charged by law with the superintendence of paving, that the work was commenced on the fifth day of October, 1859 (but six days preceding the date of the alleged notice); that immediately after the notice, he inspected the stone brought upon the ground by the contractor ; that he culled out and rejected all the stone which was not in accordance with the contract, and that such stone was then removed from the ground; that he visited the work daily for the purpose of seeing that it was properly done and in accordance with the contract ; and that after the 25th of October, Mr. Eldridge, who was then appointed General Inspector of Paving Contracts, visited the street twice a day for the same purpose. Each of these officers reports further, that, whenever he found any stone not in accordance with the contract, he threw it out and ordered the Inspector on the work to do the same. This was done to such an extent that the contractor and his friends complained to the Water Purveyor against what they termed his persecution.

The contractor was Mr. W. A. Cnmming, one of the most respectable and reliable contractors in the city.

On the 10th day of November, 1859, the Inspector in charge of the work made return under oath to the Board, that it had been completed in conformity to the specifications of the contract. The Water Purveyor subsequently certified his approval of the same, and it was thereupon sent to the Bureau of Assessors for their action thereon.

The remonstrants, if injury had been done to them by the Croton Board, had then an opportunity of appearing before the Assessors, a body entirely distinct and free from the control of the Croton Board. It appears by the papers on file that they did so,—that they were heard before the Assessors, and that the Board of Assessors decided that they had failed to make a case which would warrant them in refusing to make an assessment. It also appears by the report of the Committee on Assessments of your own Board, that they did not consider the remonstrances, attached to the assessment list by the Assessors, as required by law, sufficient to delay the confirmation of the assessment. The Board of Councilmen concurred with your Board in confirming the assessment. It was reserved for the Mayor to discover that the Water Purveyor, two Inspectors on the work, three members of the Croton Board, three Assessors, the Committees on Assessments of both Boards, and a majority of the Board of Aldermen and Councilmen, were all wrong in regard to this matter, and that the owners of property, assessed for the improvement, were right. One word as to the signatures attached to the remonstrance before the assessors. The name of J. M. Henry appears among the signatures. Mr. Henry, I am informed, is employed in the office of the Assessors, and was requested by Mr. Saxton, whose name heads the signatures to one of the remonstrances, to erase his name. Mr. Henry accordingly drew his pen through the name of Mr. Saxton, and to show that it had been stricken out by authority, wrote his own name opposite to the erasure. And yet Mr. Saxton and Mr. Henry are returned to you by the Mayor as two of the eleven owners of property who remonstrated against the assessment.*

* It is a little curious that the most prominent citizen on the list of remonstrants, has assured me, since the attempt of the Mayor to remove me, that he took much trouble to obtain an opportunity of signing one the numerous remonstrances against the proposed action of the Mayor.

SEWER IN 40TH ST., 9TH AVE., 39TH ST., ETC.

I have now to call your attention to the third charge of the Mayor. He says—

"That an error of eight inches in the maximum depth was made in the grade of a sewer now in progress in 40th street, 9th avenue, 39th street, and 8th avenue, which error may involve the city in litigation and loss, and which throws a serious responsibility on the Engineer of the Board, showing the want of the constant care and entire accuracy demanded of one to whose professional skill is entrusted the construction of important public works."

The Croton Board, intending to make a contract for this work, directed one of the City Surveyors to prepare a plan and profile of the sewer, by which the specifications for the work could be drawn and the contract let. It must be borne in mind that City Surveyors are appointed by the Common Council, and although the Croton Board has no voice in their recommendation or control over their appointment, they yet are obliged to employ them on all work which is to be paid for by assessment. The City Surveyor selected for this work was in good standing, and had been employed on similar work in common with the other City Surveyors for many years, to the entire satisfaction of the Board. He made the plan and profile, but it appears after the contract had been let and the work was progressing, that he had erred in his minutes of the depth of excavation for the sewer trench. By this error, *neither the original plan, nor line, nor the efficiency of the sewer itself was in any degree altered or impaired.* The simple effect was, that at the corner of 9th avenue and 39th street, the contractor, in order to get down to the proper grade, was obliged to excavate eight inches deeper than was indicated by the levels given on the original profile on which the contractors had based their proposals. This additional eight inches of excavation became *extra work*, for which the contractor is undoubtedly entitled to compensation. As soon as the error was discovered and reported to me, I conferred with the Corporation Counsel, and our Board have, in his opinion, and by his advice, done all that is necessary for the protection both of the contractor and the public.

Although I will not allow any one to throw upon me the
mistakes in levels, *however rare*, of any of the thirty or more
Surveyors, who are appointed by the Common Council, and
whom we are obliged by ordinance to employ, justice to the Surveyor in question has led me to make a fuller explanation on
this charge than for myself I would have made.

The fourth charge of the Mayor is as follows:—

That John Cornwell and William Conboy, employes of the
department, who had a private job for the excavation of a cellar at Yorkville, were accused before the Croton Aqueduct
Board in December, 1859, by John Chappel, and three others,
to the effect that Cornwell had furnished powder, fuse, and
tools belonging to the Department to aid in the performance of
this private contract, and had allowed men in the employ of
the Department and under his supervision to work upon this
cellar, and had fraudulently returned to the Department the
time they were thus engaged in this private contract—thus
causing the men to be paid by the Department for labor performed on private work; that notwithstanding the admission
of Conboy himself that he borrowed powder, fuse, and tools
from Cornwell, both of these persons are yet retained in the
service of the Department; that Mr. Tappen, the then acting
President, and myself, the Chief Engineer, ought to have had
them indicted for converting the public property to their own
use, and had them excluded by their conviction from ever receiving or holding any office under the city charter; that we
have upheld them in robbing the city of public property—refused to invoke the most decided vindication of the law against
them, and, by keeping them in service, have subverted the good
faith and honesty of the Department.

These are very serious charges, alleging very enormous
offences, and conveyed in very strong language, suited, if true,
to their enormity.

What are the facts? Conboy had been a laborer in a gang
of which Cornwell was foreman, and got a job, when he was not
in our employment, for excavating a small private cellar at Yorkville. His fellow-laborers, in the same gang, helped him in
the work at such intervals as they were not required by our

Department, which they had a right to do. John Chappel and three others were laborers whom we had discharged, and who, for some reason or other unknown to me, bore very ill-will against Cornwell, whom we had retained. They brought before the Board several accusations against him, to wit:—

1st. That Cornwell was interested in the job with Conboy.

2d. That Cornwell used the powder, fuse, and tools of the Department in its execution.

3d. That he returned the time of the fellow-laborers on the job as time expended in the service of the Department, and received pay accordingly; and

4th. That they thereby excavated this cellar at the cost and expense of the city.

These charges were carefully, patiently, and thoroughly investigated by our full Board, Hon. Myndert Van Schaick* in the chair. The examination took three or four days, about four hours each day. The testimony was contradictory, and failed utterly to prove the slightest fraud, or attempt at fraud, on the part of either Conboy or Cornwell. Judging from the testimony and the manner of the witnesses in giving it, the members of the Board were immediate and *unanimous* in their conviction that no possible suspicion of dishonesty could attach to the accused. This opinion, drawn from the evidence as heard and listened to, was moreover fully in accordance with the character of Cornwell, who through a steady service in the Department since its creation in 1849, had proved himself at all times, and under all trials, unquestionably entitled to our unreserved confidence.

Two facts, however, were proved, which we deemed of some importance—

1st. That Cornwell did lend to Conboy, at some unexpected emergency arising in the excavation, one keg of powder (worth about $2 75); 150 feet of fuse (worth about $1 75); and 5 or 6

* The Mayor must know that Mr. Tappen, the Asst. Commissioner, was not then acting President, as he alleges. The examination was in the early part of December, 1859, and Mr. Van Schaick did not tender his resignation until after Mayor Wood's assumption of office.

picks, the property of the city. It was proved, however, beyond all doubt, that these articles were returned, and that not a penny of loss was sustained by the city, and that nothing was further from the intention of these men than to convert the public property to their own use. The powder agent who supplies contractors from the magazine, only makes his rounds about once a week, and it is customary for contractors and gangs of workmen, who happen to be without powder, when they encounter a wet piece of rock which requires immediate attention, to borrow powder and fuse from each other.

2d. It was also proved that Cornwell had returned, on the pay roll of the Department, three or four days more time for one of the men than the man had made in the service of the city. In excuse for this mistake, it appeared that he was at that time •in charge of two gangs of men, working in different streets, and at considerable distance from each other —that this was done to save the Department the expense of an additional foreman—that in taking the time of the men on these two gangs he trusted, on the occasion above referred to, partially to a hasty look at the gangs, and partially to the report of the laborer who acted as sub-foreman during his absence. This was deemed gross negligence, and for this, as well as for lending the powder, fuse and tools, he was severely reprimanded; but, both because of his long tried fidelity, and the entire disproof of the main and very serious charges under which he was tried, it was deemed neither expedient nor justifiable to dismiss, or even suspend him from the works.

On the 25th day of February last, the Mayor sent a note asking for the papers and testimony in the case. They were promptly and cheerfully given. If looked at in the light in which the Mayor has viewed our action, is it not strange that we should have preserved so carefully such a record of it, and have been so willing at all times to submit it to the public for examination and criticism ? After reading them, he expressed surprise that Cornwell should be retained in the Department. Mr. Tappen and I immediately went to him and told him all the circumstances, and how different was the impression of the Board while listening to the evidence and looking at the persons giving it, from what it might be on merely reading the written testimony; but that if he still thought, in view of the *written* testimony, that it was injudicious to keep Cornwell, rather than that the good name of the Department should suffer, we would remove him. The Mayor replied, that under the circumstances, he wished us to let Cornwell stay for the present.

The substance of this conversation I repeated frankly to Cornwell, and advised him to see the Mayor for his more perfect satisfaction in the matter; which he said he would do, and if he could not satisfy him he would resign cheerfully rather than give us any annoyance. He called on the Mayor in company with Councilman Baulch, when, as I understood from both Cornwell and Baulch, the Mayor said he (Cornwell) had better stay where he was. Since that time the Mayor has regularly, *every fortnight, approved and countersigned the warrants* necessary for the payment of Cornwell and Conboy.

Against the laborer, Conboy, nothing whatever was proven, and of course nothing whatever was done.

A simple comparison of the charge of the Mayor in this matter, with the facts above stated, will make further comment on this point unnecessary.

SERIOUS DISAGREEMENTS AND INSUBORDINATION.

Having examined the four different accusations, with their several subdivisions, contained in the second message of the Mayor, I have now to ask your attention to his first message, in which he urges my removal upon the ground of "*serious disagreements and insubordination.*"

I am at a loss to understand how there can be any *disagreements*, between me or the Croton Board and the Mayor, in the discharge of our duties, because, by the laws and the distribution of powers confided to each, there is no subject upon which it is necessary that we should *agree*. The action and decision of the Board on the matters within their jurisdiction are final and conclusive; there cannot, therefore, be any "disagreement" between us; there may be *dissatisfaction*.

Nor can I understand how I can be guilty of "insubordination" to *him*. I am not "subordinate" to him in any sense

whatever. I am an engineer. He is a layman. I am an officer, employed professionally on the plans and execution of
public works requiring skill and science. He is the Mayor of
the city, and modestly, but justly, acknowledges his total ignorance of both. The law has committed, therefore, no such
folly as making me, in my official capacity, "subordinate" to
him ; and the act of 1849, organizing the Croton Board, and
defining the duties of the Chief Engineer, places this fact beyond a doubt.

The language and spirit of that act, an appointment for
five years, and numerous other items, show the clear intention
of the Legislature, that this most important branch of the public service should be saved, as far as possible, from the baneful
effects of the changes of party, and the demands of patronage.
The Mayor is authorized and required to see that the laws and
ordinances, as adopted, are honestly and faithfully executed by
us. And in case of fraud, or neglect on our part, he has the
power to ask for our removal, by and with the concurrence of
the Board of Aldermen, for causes to be stated. But he has no
power to suspend the operation of a law, or of an ordinance
once enacted, nor to interfere in the plans of public works under our charge, nor with the appointments or pay of the agents
whom we shall select for their execution.

He has the power, however, and is required to countersign
all warrants drawn by the Comptroller for the payment of
money from the public treasury. The Mayor may, of his own
mere will, refuse to sign warrants for the payment of workmen
legally employed.

In either case, if the Department is anxious for the dispatch
of the work, and is yet unwilling to resort to the slow and com-
pulsory process of the Courts, they are obliged to yield to the
demands of the Mayor, who thus obtains, by usurpation
a power over the Department which the law never intended
to give.

INTERVIEWS WITH THE MAYOR.

With these preliminary observations, I shall give you a
statement of my official intercourse with the Mayor, since the
first of January last, and you will then be enabled to judge
what are the facts, which, with odd confusion of ideas, he de-

nominates "serious disagreements" and "insubordination," and whether they constitute a cause for *my* removal. He has not sent to your Board any statement of the facts upon which he relies to uphold these charges, and I might, therefore, dismiss them with the above remarks; but I prefer to give the following account, for the complete understanding of the whole subject, and to enable you to form your own judgment as to the real motives of his last message.

The occasions which arose, calling for official intercourse between the Mayor and myself, were of a two-fold character.

First. The refusal of the Mayor to sign warrants for the payment of expenses in constructing certain public works, authorized by the Common Council, at one time on the ground that the ordinances were illegal, and at another time because he did not like our plan of construction. And

Secondly. An attempt to coerce the Board in the selection of their inspectors of work.

It is provided by the act entitled "An act to create the Croton Aqueduct Department in the city of New York, passed April 11, 1849," as follows:

"§ 4. The Croton Aqueduct Board, in addition to the matters charged upon them by the said amended charter, are hereby charged with the preservation of the Croton lake and waters, with the preservation of the banks of the Croton river from injury or nuisance, with the execution of such measures as may be necessary to preserve and increase the quantity of water, and keep it pure, with the management, preservation, and repairs of the dam, gates, aqueduct, high bridge, reservoirs, mains, pipes, pipe yard, and property of every description belonging to the water-works; and they shall have the construction of such new works, and the purchase and laying down of such mains and pipes as the Common Council may authorize; and also the construction, repairs, and cleansing of all the sewers and underground drains, but subject to the orders and directions of the Common Council, as to the times and places of building new sewers, and to the general plan which has been or may be adopted for the sewerage and drainage of said city. They shall be responsible for the supply of water, and the good order and security of all the works, from the Croton lake to the city, inclusive; for the exactness and durability of the structures which may be erected, and of the daily work to be performed; and for the sufficiency of the supply in the pipeyard to meet every casualty; and for the fidelity, care, and attention of all persons employed by the department in watching the works and in making constructions and repairs; and shall inspect thoroughly the interior of the aqueduct, and make the necessary repairs at least twice in each year."

*   *   *   *   *   *   *   *   *

"§ 7.   *   *   *   They shall appoint and employ all the clerks, foremen, mechanics, keepers, watchers, laborers, and other persons whom they may judge to be necessary for the performance of their duties, under this act, except the officers and clerks in the bureau of the Water Register, and shall require such bonds and securities as they may deem proper, from such of said officers and servants as they shall appoint."

It will thus be seen that, by law, the greatest responsibilities are cast upon the Croton Board, and that consequently the *plans of work*, and the *selection of agents* in their execution, are confided to that Board *exclusively*.

It had for a long time been our anxious desire to increase, as a matter of public necessity, the daily flow of water into the city. For this purpose, it was absolutely necessary to enlarge the capacity of the pipes over the High Bridge and between the two Reservoirs.

To accomplish this object at the High Bridge, without, at the same time during the construction, cutting off the flow of the water through the aqueduct, is a very delicate operation in engineering. Certain portions of the work could be done very easily by contract, but other portions it would be utterly unsafe to commit to any contractor whatever. They should be performed by the Board under the direct supervision of the Engineer and his assistants.

To ensure this necessary control over the portions referred to, we asked for, and obtained from the Common Council, on the 31st December last, an ordinance (passed by a three-fourths vote, as required by law), permitting this work to be done by the day. This ordinance or resolution was signed by Mayor *Tiemann*.

Early in the current year, it became necessary, in consequence of a change made in the grade of Eighth Avenue, to lower the main conduit-pipe in that Avenue, to conform to the new and reduced grade. The main through this Avenue supplies the daily wants of a large portion of the western part of the city. To strip this pipe, excavate a trench twelve feet deep below it, in some places through solid rock, and to lower the pipe to the bottom, *without shutting off or interrupting the entire flow of water through it, for even one minute*, was also a work too delicate to commit to the hazards of a contract. There was less danger here, than will be in some points of our work on High Bridge, but still there was sufficient risk, to make it our duty to ask for permission to do this work also by the day. A resolution to that effect was passed on the 22d day of February, 1860. This resolution was identical, *as to its legality*, with the one passed in reference to the High Bridge. This was signed on the 28th day of February by Mayor *Wood*.

On or about the 13th March, 1860, both these works were commenced, and after considerable progress had been made, and the Comptroller had signed the warrants for the payment of the workmen, we were surprised to hear that the Mayor refused to countersign them.

On the 16th May, the Croton Board, upon my motion, passed the following resolution :—

"*Whereas*, a question has arisen as to the legality of the resolution under which this Board is now proceeding with the improvements at the High Bridge, and the lowering the main pipes on the Eighth Avenue by ' days' work,' instead of by contract,

"*Resolved*, That until his Honor the Mayor decides whether he will approve of the prosecution of the work in its present form, all the said work at the High Bridge and Eighth Avenue be suspended."

On the 22d May, the Board sent the following letter to the Mayor :

"Croton Aqueduct Dep't,
New York, May 22, 1860.

" To Hon. Fernando Wood, Mayor, &c. :

The lowering of the main Croton-pipe in Eighth Avenue by days' work, in obedience to a resolution of the Common Council, passed February 28, 1860, having been suspended on the 16th inst., in consequence of some doubts expressed by your Honor as to the legality of the resolution referred to, the Croton Aqueduct Board beg leave to make the following statement and inquiry.

The lowering of the pipe became necessary, as your Honor has already been informed, by the alteration of the grade in the said Avenue. The lineal distance for which the pipe must be lowered is about the third of a mile ; the maximum vertical depth to which it must be depressed to conform to the new grade, is about twelve feet.

The daily supply of the western portion of the city makes it necessary that the flow through this pipe be entirely uninterrupted. To lower this pipe, thirty inches in diameter, when full of water, even one foot below its original bed, without breaking a joint, or without other accident, could not with any safety be done by contract. To lower it the entire distance required, (without drawing off the water,) by contract, would involve a risk amounting almost to a certainty of an accident, against the disastrous consequences of which, no amount of security received from the contractor could be set off as a compensation. It was for this reason alone, that the Croton Aqueduct Board asked that such work should be under the 'absolute control of the Engineer supervising it, so that he could at any moment either decrease or increase the extent of his operations, or stop the work altogether, as the exigen- cies of the moment might make necessary, without being influenced, as would be a contractor, by the desire, for pecuniary reasons, to drive the work to the utmost verge of safety.

It was for these reasons alone that the Croton Board asked for the authority to do the work otherwise than by contract.

When our operations were suspended, the pipe had already been stripped of its covering, a large portion of the new bed on the lower grade prepared by excavation, &c., and the pipe supported above it by the screws and other appur- tenances necessary for the lowering. The process of lowering had also been commenced. The pipe now stands on temporary supports, at some distance above the surface of the Avenue; and although every precaution and care have been taken to render it secure and prevent any accident, it is by no means in a position of absolute safety.

We would further beg leave to say, that in the contract made by the Street Commissioner for grading said Avenue, a specified time was reserved, during which this Department should have a right to the possession of this portion of the Avenue, for the purpose of lowering the pipe. No more of that time now remains to us than is necessary for the work we have to perform.

Under these circumstances, we have deemed it our duty to make this state- ment, and to ask whether in your opinion the necessity of the case requires that the work in question should be resumed, and whether in the event of its resumption, you will sign the warrants for the money required in its prosecution.

Very respectfully,
Your obedient servant,
(For the Board.) THOS. STEPHENS, President."

On the 23d May, a letter was received from the Mayor in reference to the work on the 8th Avenue, as follows,—

MAYOR'S OFFICE,
New York, May 23, 1860.
"THOMAS STEPHENS, ESQ, PRES'T CROTON AQUEDUCT DEP'T.

SIR,—In reply to the communication from the Croton Aqueduct Department of the 22d inst., I beg leave respectfully to say, that for the reasons stated—as well as that in my judgment the proposed lowering of the main Croton pipe in Eighth Avenue *by days' work, does not conflict with the city charter*—I have no objection to the renewal of that work forthwith.

Very respectfully,
FERNANDO WOOD, Mayor."

On the 31st May the Board again addressed to the Mayor another urgent request to countersign the warrants for the High Bridge, as follows,—

CROTON AQUEDUCT DEP'T,
May 31, 1860.

"To His Honor, Fernando Wood, Mayor, &c.:

Sir,—The work which was in progress for the improvement of the High Bridge was suspended at the same time the operations in lowering the main pipe in 8th Avenue were suspended, and for the same reason, viz., the question raised as to the legality of the resolution under which we were acting.

This resolution was passed by the last Common Council, on a representation by this Board of the great risk which would be incurred in doing the necessary work by contract. The great mass of the materials to be supplied, and a large portion of the mechanical part of the work can be done by contract; but a part of it cannot be so done, without danger as imminent as that set forth in our communication to you as to the work on 8th Avenue, dated May 22d. Should an accident occur on High Bridge, the supply of water would be cut off, not solely from a portion of our streets, but from the entire city.

Under these circumstances we beg leave to set before you this urgent expression of opinion that, so far as the safety of the city requires it, this work should be done by the day.

This embraces but a very small portion of the entire work, and we beg to ask whether, in view of the necessity, which urged our Board to ask for the resolution above referred to, you will countersign the requisitions drawn for such expenses as may be incurred in prosecuting otherwise than by contract, such portions of the work as, in the judgment of this Board, it is utterly unsafe to allow to be done by contract.   Very respectfully,

Your ob't servants, &c."

It will thus be seen that down to the date of this letter, the Mayor had undertaken to determine (as in the case of Cornwell and Conboy), whom we should retain in our employment, and whom we should dismiss, and (as in the case of the High Bridge and that of the 8th Avenue) what work we should do and what work we should not do; although both involved the same points of legality. All these questions were, as I have shown, placed, by the law, exclusively within the jurisdiction of our Board. We were, however, compelled to submit, because, without money to pay the men, no work could be done.

But the Mayor made further demands; he notified the Board that he required that all Inspectors appointed by us should be sworn before him at his office. This course was unusual, had never before been adopted, and was not required by the Act of 1849;—but wishing to interpose no objection, so far as the mere ministerial administration of the oath was concerned, we directed the Inspectors to attend upon him for that purpose.

About this time—the last week in May—the rumor prevailed in the city and in the public prints that the Mayor intended to remove all the Heads of Departments, myself included.

When I heard the rumors I called upon him, and for reasons which will suggest themselves to your minds, I beg leave to state our conversations somewhat minutely.

I told him frankly, that perhaps my calling might be considered rather abrupt under the circumstances, but as I was generally in the habit of going at my object in a straightforward way, I could not help doing so in this case. First telling him what I had heard, I said that, as I presumed the only good grounds for removing an officer were either the want of ability or fidelity, and as I had been flattered, by the general voice of the community, into the belief that the first charge *would* not *probably* be the one, and as I was conscious that the second *could not possibly* be brought against me, I presumed he must be under some misapprehension or misinformation in regard to facts, and I therefore called to ask him if such were the case, and to say that I was ready to answer and give explanations upon any points required.

He said that *he had heard* of *nothing* and *knew nothing against me*,* and had not yet told any one he intended to remove me. I told him that my object was not to ascertain what he had told any one, but what he intended to do. He then said, in so many words, that he had not arrived at his present position without a severe struggle, and without incurring many obligations; and that *his object in removing heads of Departments was to get control of the Departments, so that he could put in those who would co-operate with him, and also could pay off his obligations.*

• Some other conversation ensued, which was mere amplification of this. As I was going out, the Mayor told me that he had not yet made up his mind what he would do, but that he would not act at all in my case, until he had seen me again.

* Please to observe that this interview was on or about the 26th of May; that he was in possession of the facts concerning the Reservoir on the 3d of May; and of the testimony taken in reference to John Cornwell on the 25th February.

A few days after this he sent for me. When I came to his office, he told me it was said that an arrangement had been made by me with the Board of Aldermen, by which I was guaranteed my place, on condition that they should have all the patronage. I replied that it was out of my power to prevent such calumny; and that no friend of mine would believe me capable of doing anything of the kind.

He said it was an Alderman who was friendly to me that had made the statement "that I did not want the patronage, and had said I wished it were in my power to get clear of it." I answered, that so far as patronage *per se* was concerned, it was very true that I did not want it, for even if I had friends to provide for, all personal feeling would be entirely secondary to my duty to the public as Engineer, which required me to select foremen for their merits alone ; that the pressure of politicians upon me in behalf of their friends was so great for every place, however insignificant, that the selection of inspectors was one of the most wearing and unsatisfactory labors of my office ; and that, if I could do so conscientiously, I would willingly let Aldermen or any other person select for me ; but that, while I was willing and anxious to accommodate the Common Council by the employment of their friends to the utmost of my power, my regard for my professional reputation, and for my duty, would always forbid my yielding to any one uncontrolled power to appoint inspectors over work for which *I was* solely *responsible*.

That I never had, and never should, forget myself so far as to make any such agreement with any body of men, or with *any single man*—that strong as was my desire to retain my position until the Reservoir was finished, (it being a work which could not but prove a great credit, professionally and publicly, to all who were engaged upon it,) the desire was not so strong as to make me forget my duty as a man and an officer, and that I would not hold my place at the expense of any *such breach of trust as that implied in his remarks*. The conversation then ended, and I withdrew.

A few days after this interview, I was informed that the Mayor refused to swear in an inspector whom the Croton Board had appointed to supervise the construction of a sewer in the 6th Avenue, between 50th and 51st streets, by contract.

This refusal made it quite evident that the order of the Mayor before mentioned, requiring all Inspectors to be sworn before him, was intended to include something more than the mere ministerial administration of the oath ; yet we submitted.

On the 2d of June, 1860, the subject of this refusal, and the continued suspension of the work at the High Bridge, came under the consideration of the Croton Aqueduct Board; which, on my motion, adopted the two following resolutions :

(Extract from minutes Croton Board.)

" *Resolved*, That J. J. Noble be suspended until the Mayor shall have time to investigate certain charges alleged against him.

" *Resolved*, That a communication be sent to the Mayor, stating the necessity for going on with the work at High Bridge, and asking whether he will countersign the warrants therefor."

On the ninth day of June, the Mayor sent to our Board the following letter, exculpating the Inspector :

MAYOR'S OFFICE,
New York, June 9, 1860.

" I am satisfied that the charges against J. J. Noble are unfounded.

FERNANDO WOOD."

At the time this letter was received, I was absent, owing to illness, but I am informed by my colleagues that Noble was sent back again *to the sewer* on which he was first appointed ; but that the Mayor refusing to recognize him on that sewer, he was *again withdrawn.*

On the 20th June, the Board again addressed another letter to the Mayor in regard to the High Bridge, as follows,—

CROTON AQUEDUCT DEPARTMENT,
June 20th, 1860.

" To His HONOR FERNANDO WOOD, Mayor, &c.:

Sir,—The Croton Aqueduct Board beg leave respectfully to call your attention to their communication of May 31st, relative to the proposed improvement of High Bridge.

Copies of opinions of the Corporation Counsel, given in reply to a question submitted by one of the Committees of the Board of Aldermen, and also to a question submitted by this Board, are herewith furnished ; the points involved being similar to those raised in the matter above referred to.

The work of enlargement on High Bridge was authorized to be done by days, work on the 31st day of December, 1859.

‿‿ The work preparatory to the construction of a new pipe was commenced on the 13th day of March, 1860. The work was suspended, at the request of the Mayor and by order of the Croton Aqueduct Board, on the 16th May, 1860. The gravel covering of the Bridge and pipes had at that time been removed, and the dry wall around the Gate-house on the Manhattan side had been taken up, preparatory to making a connection with the Gate-house.

The Bridge is now, and has been since operations on it were suspended, in this exposed condition, and it is most important that some definite conclusion in regard to it should be arrived at, without further delay.

If the work is to be abandoned, the covering mass and material which has been removed from the pipes and masonry should be replaced as a protection. If it is to be prosecuted, no time is to be lost. The delay already experienced will make it necessary to cover with a temporary roof, extending the whole length of the Bridge, work, which must now be unavoidably left unfinished. Further delay will render it impossible to get our pipe-work in such a state of forwardness as to be available during the coming winter. During the cold weather of last winter, the water in the lower Reservoir was reduced to so low an ebb, as to create serious apprehensions of the result of any large fire which might have occurred at the time. During the coming season the increased population of the city will superinduce an increased consumption of water, and the difficulties and risks heretofore encountered will be proportionably magnified, unless the proposed improvements be completed. These improvements are the enlargement at the High Bridge, and the additional connection by a four-foot pipe between the Receiving and Distributing Reservoirs. Both these works are now suspended, and we beg leave to report that some action should be taken at once, if the surplus water now running waste at the north end of the High Bridge, is to be made available for the comfort and safety of our citizens and their property.

The greater part of the work at both the points above named, can be done by contract; but inasmuch as certain parts of it cannot with safety be done otherwise than by days' work, application was made to the Common Council, which resulted in the ordinances above referred to. Should you decide that, under the circumstances, you will sign the warrants necessary for the payments on the work, we can go on, and yet be prepared for the coming winter. If not, and we are thrown upon the delays and dangers of the contract system, the difficulties may be very serious. The Board feel satisfied that a personal inspection of the site of the proposed work, would aid in the removal of the objections expressed by your Honor, and they will be happy, if opportunity be afforded them, to accompany your Honor to the same, at any time you will be pleased to name as consistent with your other engagements.

Very respectfully,

Your obed't servants,

THOS. STEPHENS,

THOS. B. TAPPEN,

A. W. CRAVEN,

Croton Aqueduct Board."

On the 28th June, having received no reply from the Mayor, I called upon him at his office, when the following conversation occurred :—

I asked him if he had yet made up his mind whether he would countersign the warrants for the payments on the work, and told him that, as had been fully set forth in our two communications to him, dated May 31 and June 20th, the necessity for some action on the subject was absolute and immediate.

He answered that he had made up his mind not to sign the warrants. I asked him if he would notify our Board in writing of his determination, and he said he would do so the next day. I then observed that he was assuming the responsibility of delaying a work, the immediate progress of which was most important to the city, both as to comfort and to safety. He said he should be governed by his own judgment in that respect, and added that his reason for opposing the work was *that he did not like the plan.* He did not state in what respect, nor did I ask him. I answered that with all due deference to him, that the plan was a matter committed by the Legislature to the judgment of the Croton Aqueduct Board. He asked if I meant to say that he had no right to any opinion, or no right to supervise the Departments. I answered that as chief magistrate he had not only a right, but that it was his duty to inquire into the manner of conducting business *in all the Departments,* and to check any dishonesty or impropriety of conduct, if he found it; but that in all matters which, by the charter, were exclusively placed under the control of the different departments, as to specifications and plans, he most surely could claim no prohibitory voice; and particularly in a matter of engineering science connected with the Croton Aqueduct, he was not the officer legally appointed to decide; and that he could not properly make his opinion a good ground for stopping a work of such pressing importance, and in regard to the plans of which, the proper Department had fully decided. He said he should undertake it. I then repeated my request for his decision in writing, and he repeated his promise to let us have it without delay.*

---

*He has not done so to this day.

I then spoke to him about the Inspector on the sewer in 6th avenue, whom he had refused to recognize because of certain charges alleged against him by the contractor on said sewer, which charges he, the Mayor, had, in writing, notified our Board were unfounded. He said he would recognize him *on any other sewer* than the one for which he had been appointed. I then urged upon him, that *if a contractor could select his own Inspector*, and *be upheld by the Mayor* in doing so, there would be little use in appointing Inspectors at all; that such a course would establish a precedent entirely subversive of all power to have public work done with any degree of honesty, or fidelity to the city. As he would not be changed by any arguments, I withdrew.

On the 6th of July, I offered, at the meeting of the Croton Board, the following preamble and resolution :—

"*Whereas,* On the 3d of May last, an Inspector was selected by this Board to superintend the sewer about to be constructed in Sixth Avenue between Fiftieth and Fifty-first Streets ; and

Whereas, His Honor the Mayor (who had already claimed that all inspectors under this Department should be sworn in by him) refused to swear in said Inspector, in consequence of certain charges which he stated had been alleged against him ; and

Whereas, On motion of A. W. Craven, Chief Engineer, said Inspector was suspended from the work, in order to give his Honor the Mayor an opportunity of investigating the said charges, and the Mayor was advised of said suspension ; and

Whereas, By a letter dated June 9th, the Mayor informed this Board that the charges against said Inspector were disproven, but notwithstanding said letter and said exculpation, his Honor still refused to swear him in, giving to said Inspector, as a reason for such refusal, that he had promised the contractor on said work not to do so ;

And Whereas, acquiescence in such a reason for the dismissal of any Inspector, would lead to a total subversion of the ability of this Board to have any of the work committed to its charge performed honestly, and according to the contract made in behalf of the parties to be assessed ; therefore

Resolved, That said Inspector be re-appointed on said sewer in Sixth Avenue, between Fiftieth and Fifty-first Street, according to the original selection of this Board ; said appointment being by virtue of the authority given by the charter to this Department, to appoint Inspectors on all work for the faithful performance of which this Department is by the same charter *alone held responsible.*

On the 11th July he sent his first message to the Board of Aldermen, removing me from office for " serious disagreements " and " insubordination."

On the 18th July he sent a second message, removing me for additional reasons, which I have stated and examined in the first part of this answer.

And you, and my fellow citizens at large, will probably be surprised to learn, that though I have been at all times ready and willing to make the statement and explanations I have now presented to you, and although the alleged reasons of the Mayor contain the most serious charges against my character, both personal and professional, yet that the Mayor never alluded to them in any of my interviews with him, nor did I ever hear of them *until I read his charges in the newspapers which carried them into all parts of the United States.*

CONCLUSION.

I dismiss this subject to your final determination, with a brief statement of the conclusions clearly deducible from the facts.

1st. It is not true that I have made any "changes" whatever in "the plan" of the Reservoir, either "extensive" or "peculiar," or otherwise. No work *has been done, or is contemplated* upon the Reservoir except such as is specifically mentioned and provided for *in the contract,* and *at prices therein limited.*

2d. It is true that there are variations in the *quantities* of work and materials, as originally estimated, and as actually to be done.

3d. It is not true that these variations are "remarkable," except in one sense, that, considering the vastness of the area, and the impossibility of making an *exact* and subterranean estimate, they are "remarkable" in presenting comparatively so slight a difference.

The area is 106. acres, presenting an irregular surface of rocks and morass, the rocky hillocks rising in some places thirty-five feet above the swamps in the same area.

The depth of our excavation exceeded, in some places, forty feet.

The estimate was of necessity made before the surface was broken for the work, and the *quantities*, and particularly the *qualities* of the various earths and rocks to be found beneath the surface were, as every engineer knows, and as I have explained to you, matters more or less of uncertainty.

4th. It is true that for these variations, requiring *additional* work and materials, the contractors will be entitled to some $100,000 beyond their bid of $632,743, which was based on the quantities and qualities stated in the estimate; but it is not true that the cost to the city will be increased that amount beyond *my estimate of the cost*, made for the use of the Government. Such estimate was (exclusive of the gate-houses) $1,025,000. The Reservoir will be built for $150,000 less, at least; probably much lower.

5th. It is not true that in "every case" in which these variations have occurred, "they have proved very beneficial to the contractors." On the contrary, several of them are done at prices which indicate a loss, and the others are done at prices barely remunerative, and which a majority of the competitors were not willing to do for fifty per cent. advance.

6th. It is not true that I made any "mistake" in the award of the contract.

I. *I* did not make the award: it was made by Messrs. Van Schaick, De Forrest, myself (as the Croton Board), and the Hon. A. C. Flagg, the Comptroller, unanimously. Fairchild & Co., to whom the award was made, were the lowest bidders.

II. Even had I alone made the award, it was made *strictly* according to the requirements of the charter; there was no *mistake in any one point of the letting*.

I may be excused from saying here, that notwithstanding the difficulty attendant upon explanations on professional subjects, I have clearly shown that the charges of the Mayor which were set forth with such minute detail of figures and items, are utterly unfounded when examined in the light of intelligence and truth. In the whole conduct of the work on the Reservoir, from the first examination of its site until the present moment, that course has been adopted, and those precautions taken, which the experienced and skillful men in my profession would, under similar circumstances, suggest and advise ; and I hope you will be as much surprised as my professional brethren are, that the very facts which I would invoke as evidence of capacity, integrity, and judgment, are perversely distorted so as to assume, to the eye of those who do not understand the subject, the aspect of folly, ignorance, or something worse.

In providing, by numerous specifications, for contingencies dependent upon the developments of our operations, and in availing myself of those provisions, I have done only what was absolutely required by my duty as an officer, to ensure the perfection of the Reservoir. Had I done otherwise, I could not have, as now, the right to hope that with this great public work my name may hereafter be honorably associated.

7th. It is not true that the paving in Forty-ninth street was executed without " proper vigilance " on my part. The Water Purveyor, charged by. law with such work, examined and passed it. The Croton Board, the Board of Assessors, the Board of Aldermen, and the Board of Councilmen confirmed it. Against the unanimous action of these five separate bodies, there is no evidence but the unsworn statement of nine property-holders, interested to defeat the assessment for the work.

8th. It is true that an error of eight inches, in the maximum, was made in the grade of a sewer in Fortieth street ; but it is not true that the error was made by me, or that I am responsible for it, or that the city will suffer any loss by it. The error was made by a city surveyor, *appointed by the Common Council*, and will simply involve the additional charge by the contractor for extra excavation, averaging four inches in depth, for a limited distance.

9th. It is not true that any of the men in our department have been guilty of the fraud of converting the public property to their own use, or that we have retained any such men in our employment.

4

Some discharged laborers got up a prosecution against
Cornwell, a foreman of a gang (employed by us for eleven
years with entire satisfaction), charging him with gross offences.
We examined witnesses patiently for four days, and the entire
Board, Messrs.Van Schaick, Tappen, and myself, unanimously
dismissed the charge. We were the proper tribunal before
which to try the case, and we did try it thoroughly and honestly.
This the Mayor has in effect admitted ; for since the explana-
tions given him, he has signed the warrants for the payment of
Cornwell and Conboy semi-monthly, up to the present time.
So much for the second message. Now for the first.

10th. It is not true that there have been any " disagreements," serious or otherwise, between the Mayor and myself.
Such a state of facts is simply impossible ; because, by law, in
the discharge of our respective duties, there is no one subject
upon which it is necessary that we should *agree*.

11th. It is not true that I have been guilty of " insubordination" to him. I am an *Engineer*, charged with the responsibility of the plan and execution of public work requiring skill
and science. He is a *layman*, and the law is not guilty of the
absurdity of making me " subordinate " to him in the discharge
of my duty. On the contrary, the act of 1849, organizing the
Department, imposed upon me a personal responsibility, and
makes my decision, within the sphere of my jurisdiction, final
and *exclusive*.

12th. But it is true that after his repeated attempts to usurp
the power of the Department, and to make it the slave of his
will, I have raised those objections which I was legally entitled
to interpose, and which are demanded of me by every consideration of fidelity. to the city.

By *abusing* the power confided to him of countersigning
the warrants of the Comptroller for the pay of our men ; granting or refusing his signature at pleasure ; *he has undertaken
to say whom we shall retain in our employment, and whom we
shall not—whom we shall select as Inspectors, and whom we shall
not—and what work, authorized by law and the ordinances, we
shall execute or omit.*

He has arrested the execution of two very important public works, in the very height of their progress ; and, by his action on one of them, defeated or delayed the long cherished desire of the Department to increase the supply of water. One of these works, that on Eighth avenue, was in course of execution under an ordinance of the Common Council, *signed by himself;* and the other, that on the High Bridge, under an ordinance signed by *his predecessor.* Both of these were suspended upon his asserted doubts of their *legality.* He withdraws his objection to the first and least important of the two, and grants us *his permission* to proceed.

He refuses *his permission* to the latter, involving the *same* point of *legality,* and starts an objection to the *plan* of the work—a matter wholly within the province of our Board, and wholly upon *my* responsibility.

For the sake of not having any important public work delayed, I have never, for one moment, stood upon the question of subordination ; but have yielded in every point until it came to a question involving my fidelity as an officer. He refused to countersign the warrant for the pay of an Inspector whom we had appointed upon a sewer, upon the ground *that he had promised not to do so, at the request of the contractor* whose work the Inspector was to supervise—expressing his willingness to recognize such Inspector *upon any other sewer.* I immediately introduced resolutions into our Board, pledging my colleagues to resist these usurpations ; and within five days afterward he sent to the Board of Aldermen his first message, proposing to remove me for " serious disagreements and insubordination."

Respectfully submitted.

A. W. CRAVEN,
Chief Engineer Croton Aqueduct Department.

City and County of New York, ss :

Alfred W. Craven, of said City, being duly sworn, says that the foregoing answer is true to the knowledge of this deponent, excepting that part of it which refers to the instrumental examinations of the assistant Engineers of this deponent, and as to those matters, deponent believes it to be true.

A. W. CRAVEN.

Sworn to before me, this 1st.
day of September, 1860.

EDWARDS PIERREPONT,

Justice Superior Court.

www.ingramcontent.com/pod-product-compliance
Lightning Source LLC
Chambersburg PA
CBHW031806090426
42739CB00008B/1189